RESCUED BUILDINGS

RESCUED BUILDINGS

The art of living in former schoolhouses, skating rinks, fire stations, churches, barns, summer camps and cabooses.

Roland Jacopetti
Ben VanMeter
Photographs by Wayne McCall

CAPRA PRESS • 1977 • Santa Barbara

Library of Congress Cataloging in Publication Data

Jacopetti, Roland, 1937-
Rescued buildings.
1. Buildings—Repair and reconstruction.
2. Dwellings. I. VanMeter, Ben, 1942- joint author.
II. McCall, Wayne, joint author. III. Title.
TH3411.J3 690'.8 77-21399
ISBN 0-88496-110-9

Acknowledgements: Design and production, Marcia Burtt; typesetting, Camera Ready Composition and Foster & Horton; camera work, Santa Barbara Photoengraving and Haagen Printing Co.; presswork by Haagen; binding by Area Trade Bindery, Los Angeles.

Photo credits: Nipton Troll, page 8, by the Los Angeles Times; historic photographs of Bloomfield Schoolhouse, pages 17-18, by Charlie Pierce; Mowry's Opera House photo from California Historical Society Library, page 26; photographs of Max's Castle, pages 28-29, and Spring Hill, pages 34-35, by Roland Jacopetti; Lighthouse, page 31, by Ben VanMeter; historic photographs of Sunshine Camp, pages 80 and 82, from the collection of Roland Jacopetti; photograph of the authors, page 96, by Susan C. McCall.

Capra Press
631 State Street
Santa Barbara, CA 93101

CONTENTS

6 Foreword
8 Nipton Troll
9 Mitchell Road Barn
10 Water Tower
12 Wine Barrel House
14 French School
15 Bodega School
16 Marshall School
17 Bloomfield Schoolhouse
24 Olema School
26 Mowry's Opera House
28 Max's Castle
30 Reno Hotel
31 Lighthouse
32 Silo
33 Roost
34 Spring Hill
36 Bodega Creamery
39 Venice in America
48 Trains and Such
49 Train Station
51 Parlor Car
52 Caboose Couplet
54 Trolley
55 Caboose
56 Mitchell Road Stable
59 San Francisco Church
61 Seneca Falls Church
62 Silver City Church
63 Druids Hall
65 Sandcastle Gallery
66 Oak Barrel Triptych
68 The Water Barrel
69 Palomar Skating Rink
71 Service Station
72 Chemical #5
74 Last Resorts
76 Camp Imelda
79 Sunshine Camp
93 Afterword

FOREWORD

The past decade has spawned thousands of books concerned with the revolutions in lifestyle which have lately struck our civilization, and the challenges to human survival endemic to life in the 20th century. This book, though not intended to be a survival tool, comes from those of us caught in the joy of living in buildings rescued from our abandoned civilization. We hope the satisfactions rub off.

Civilization, as we know it, is ending all the time—businesses turning belly up, institutions becoming obsolete and moribund, economic and social customs fluctuating whimsically and wreaking havoc on established modes of behavior. Our book investigates the virtues of recycling outmoded properties into home-sweet-homes. We hope to encourage readers to inject some energetic imagination into the serious business of providing shelter for themselves. Moreover, as characterful old buildings are torn down to make room for parking lots and architectural excreta, we are making a bid to save these venerable white elephants from the bulldozer's tooth.

Most houses, condos and apartments are inconvenient for people who are involved in home crafts, run home businesses or who feel cramped by conventional dimensions. Victorian residences, for example, are charming places visually, but designed for compartmented living, one room at a time. Space is squandered in elaborate systems of walls and doors to insure that rooms' functions don't get mixed. Modern houses aren't much better. Generally, they're designed to fatten builders' profits while providing occupants with an illusion of gracious living beamed straight from a 19-inch color TV. No matter how carefully calculated for traffic flow and space utilization, they usually prove boring and inconvenient, and lack warmth.

What makes living in a rescued building so much better? This book will tell you and show you. Each building has its own quality, its own charming or outrageous uniqueness affecting the lives within—like the old shoe store built to endure heavy foot traffic with its solid doors, industrial hinges, double-thick windows, thus giving its resident a sense of security and duration. Or the auto repair garage converted by a family who focused their lives around the grease pit they transformed into a congenial hot tub.

How people adapt unusual living spaces to fit their needs (and vice versa) is much of what this book is about. When an architect designs a house he usually scales it to human occupants and uses day-to-day human living as the key to design. Converting a gas station or firehouse, on the other hand, poses a different problem. The general esthetic and function is already there, and the occupant is challenged to establish new scale and use. Taking on this challenge can certainly open windows in your preconceptions of what home is all about.

As you will see, advantages and problems go hand in hand. Recycled buildings may offer magnitudes of space to raise seven kids, hang giant canvases, grow a multitude of house plants, stand back and hurl epoxy missiles at expanses of masonite, or weld together a dozen '49 Buick fenders, but conversely one can easily be overwhelmed by sheer size. What to tear out and what to keep, what to build and how big? You face the same dilemmas whether you buy ten acres of rolling hillside or a 50-by-150 foot laundromat. How can we partition some of it, how does the plumbing work and, above all, how do we heat it? The inhabitant of a rescued building must be guided by an inner light.

That shoe store, for instance, was designed to house employees, customers, and a few thousand pairs of shoes without feeling crowded. Strip out the fixtures and shoe boxes and you suddenly feel very, very small. You have to walk farther and talk louder, or huddle with your loved ones in an alcove. Most people we talked to found they needed people space as well as work space, places to feel warm and cosy in times of relaxation.

At the other extreme we found people living in small spaces like redwood wine vats, where miniaturization and efficiency become essential. As in ship and trailer living, nothing is wasted—they had to put a single area to many uses and exercise ingenuity to obtain privacy. But adapting your life to minimal space may also provide enhancement. You must organize out of sheer necessity—become efficient or move out.

Building recycling is more than just an isolated phenomenon, and of greater scope and social importance than simple lifestyle enhancement. In September, 1976, the Los Angeles *Times* ran an article headed "Abandoned Homes: Idle U.S. Resource." The cities of America, claimed the article, are filled with hundreds of thousands of abandoned buildings, mostly awaiting demolition. These constitute a major national resource going completely to waste, as most could be made attractive and useful with only minor repairs, at a cost of roughly half that required to build new structures. Yet, with millions of Americans living in substandard quarters and the cost of construction rising every month, the buildings remain empty—another aspect of our no-deposit-no-return culture, as well as an insight into the wasteful way cities grow. City centers become deserted as the live population retreats farther into the suburbs to escape overcrowding, crime, noise and pollution. The well-to-do who stay in the urban center barricade themselves into high rises protected by armed guards and alarm systems. The poor are given the task of subsisting in ghettos. The rest is abandoned.

In New York City alone, HUD estimates 35,000 housing units are abandoned each year. And they should know—because at one time HUD was in possession of more than 100,000 abandoned homes acquired by foreclosure on federally guaranteed loans. If grief smelled like rotten eggs, those foreclosures would stink up the entire Northern Hemisphere.

Not just houses are being vacated. Office buildings, supermarkets and factories are being deserted and stand vacant for years until the wrecking ball puts them out of their misery. A Rutgers University study finds the probability of abandonment increases when the structure is owned by an institution or other absentee landlord. A HUD spokesperson believes over a third of America's housing needs could be met by restoration and recycling.

Some American cities are already taking action. Boston is spending millions to convert a hundred acres of waterfront property into a living and shopping complex. Manhattan has used tax incentives to promote the conversion of numerous empty hotels and office buildings into apartments. In Los Angeles a plan to transform the abandoned headquarters of the Pacific Telephone Company into a shopping and residential complex for senior citizens fell through because potential investors felt the downtown location wasn't a good environment for old folks. Maybe not, but lots of them have to live there anyway because they can't afford rents elsewhere.

So don't wait around for the movie. Read the book, join the fun, rescue a building, and rearrange the world.

Roland Jacopetti / Wayne McCall / Ben VanMeter

From western desert to upstate New York, from the humble culverts of the Nipton Troll to the magnificence of the Mitchell Road barn, the following pages portray what lies between these extremes in the way of rescued buildings.

THE NIPTON TROLL

He's known as the Nipton Troll. He lives scot free in three rooms, each with running water when it rains. One is a storeroom for food, water and firewood; another is a bedroom with America's deepest wardrobe; and the third is a library/reading room stocked with magazines, newspapers and books. Each room is tubular, four feet in diameter and fifty feet long. His laundry and bathroom are two miles away at a Caltrans rest area. His backyard stretches long miles through the desert between Los Angeles and Las Vegas. With the troll in his culverts, we begin our book. Other living places we portray may be more grandiose, but none more serene.

MITCHELL ROAD BARN

A fine old barn, adorned with classical trim and a cupola, has been transformed into a magnificent country home: where cows were once milked, a dining table stands. A central column stairwell leads to the loft. The second-generation rescuer of this building said, "Mother lived in a chicken coop. Suburbia was the last place she was willing to live."

WATER TOWER

A landmark in South Pasadena for three-quarters of a century, this shingled water tower was converted into a dwelling decades ago. It sits next to the Victorian home it served atop a hill, surrounded by a throng of modern crackerbox houses—like a castle stronghold of old amid the peasantry. The bottom story is the living area; kitchen and bath form the second; and the bedroom is the airy third floor.

WINE BARREL HOUSE

One fine day, Gordon and Shirley Sanford drove to Ukiah, California (spell it backwards) and bought a gigantic redwood wine barrel from a winery that was converting to aluminum aging vats. They set the barrel upright on a coastal hillside, roofed it, and added doors and windows, including some of Shirley's original stained glass. They installed a floor halfway up the barrel, creating a second story, purchased a smaller barrel to transform into a connecting bathroom, then plumbed and wired. The result—a graceful and comfortable small dwelling at a cost of less than two thousand dollars. The twenty-six-thousand-gallon container measures eighteen feet in diameter, a bit over five hundred square feet on two floors. On foggy winter nights when wind blows from the ocean and a fire roars in the grate, the walls still exude a faint odor of the grape.

This freestanding stairway leads to the second story bedroom. The bath is an adjoining small barrel.

Clerestory windows give the bedroom a round view of coastal hillside and ocean.

The Sanfords removed staves and inserted stained glass windows in the interstices.

FRENCH ROAD SCHOOLHOUSE

A traditional school becomes home. The interior of this one-room schoolhouse has been converted from a large undifferentiated space into a Victorian maze of small rooms.

BODEGA SCHOOL

Bodega's big, four-room schoolhouse, with its curious quasi-classical cupola, was built in 1873 for the children of what was then a burgeoning redwood boom town. The surrounding hills were covered with immense first growth redwoods. When San Francisco burned and had to be rebuilt, almost all these ancient trees were clearcut all the way north to Mendocino County. San Francisco rose from the ashes on redwood beams and siding. Logging communities like Bodega flourished, then declined as stand after stand of the largest living things were decimated. John Muir said, "As timber, the Redwood was just too good to last."

The school was closed by the county in 1962 and remained empty and unwanted for five years until Mary Thames and Tom Taylor bought it and converted it into an art gallery with an elegant upstairs living space.

Bodega School nestles in northern California pastureland.

MARSHALL SCHOOL

Al Clarke fell in love with Marshall School at first sight. Built in 1876, it was in continuous use as a school until 1965. Soon after it closed, the proprietary townsfolk liberated its artifacts, including its desks and, yes, its bell. No one saw the bell for three years until it suddenly materialized in front of the local Catholic church.

Clarke carefully preserved the old building, including the last lesson chalked on the blackboard. He strengthened the upper floor for use as a bedroom, converted the bell tower

into a meditation chamber, and installed a spiral staircase made from hatch covers found in nearby Tomales Bay.

When a small rural community atrophies or loses its school to consolidation, often local alumni still cannot see it as anything but "The School." It takes an outsider to see it for what it really is: a noble old building in need of repair, and with the potential of becoming a unique and comfortable place for an interesting person to work and live. What better fate for an old schoolhouse than to have an aerospace engineer-jeweler-sculptor-woodworker fall in love with it?

BLOOMFIELD SCHOOLHOUSE

"Retired Schoolhouse, 2¼ ac., Sonoma County"
Real Estate Want Ad, *S.F. Chronicle*

Ever since Sandra and I lived in the central city ghetto loft of Mowry's Opera House, we had dreamed of having a similar space in the country, surrounded by rolling hills and clear sky. Something "like a loft" but with room to dance, paint, make films, raise vegetables, and run ponies and kids. Our first view of Bloomfield School was glorious. It offered everything we wanted: two big rooms, each 30 feet square (one for house and one for studio), kitchen, lunchroom, boys' and girls' bathrooms, cloakrooms, and a generous attic that could accommodate two bedrooms. And all this was surrounded by two acres of rich rolling land. The asking price was $38,000. I glanced at my checkbook and saw I had exactly $38. I felt this a good omen, if I could only add three zeros. Friends, to our everlasting gratitude, lent us enough for a down payment, another wealthy friend in Hawaii redeemed our sub-par credit by co-signing and thus we survived a month of outrageous torture known as escrow.

It was pouring rain the day we moved in. I carried Sandra over the threshold while a procession of six friends followed with boxes and goods. We managed to scrape up a celebratory meal for everyone and after they left, we settled down to relax and unpack. The kids, Lila and Benjamin, began searching the boxes for their toys. Sandra and I slowly let the reality sink in—our dream had come true. Suddenly, "BAM! BAM! BAM!" Someone was pounding the front door. Sandra opened it and there stood 200 pounds of agitated shitkicker, our new neighbor. He was wearing knee-high cowshit-encrusted rubber boots, a hunter's vest and a baseball cap.

His welcoming words were, "There's a tree on your property that hangs over my calf lot and's gonna fall down in the first storm. You've gotta get a tree removal company to take it down so it falls on your side of the fence." We went out to take a look at the tree, one of a dozen giant cypresses, a half century old, planted as a border to the school property. The base of one tree had been ravaged by termites and leaned precariously over the fence, aimed diagonally across the calf lot towards a house. I asked if he lived there.

Proudly erected in 1921-22 to prepare boys and girls for "20th Century living."

"Nope." He spat on the ground. "Down the hill."

"Anybody live there?"

"Nope, but it belongs to me. I've got these hundred and five acres and I lease the forty behind you with an option to buy." He spat again.

"Think it'll reach the house if it falls?"

"Might." He shrugged and spat. I could see no calves.

"Bloomfield Follies" presented by students of another era.

Closing day, 1966.

"Well," I began, "it was the hardest thing I've ever done just getting here, and I surely don't have five hundred dollars to have someone pull that tree over this way. If you could just keep your calves out of here for awhile, I'll take care of it soon as I can."

He actually smiled a little and asked if I was a college professor. Said he'd always been a dairyman and had this place for five years. He went on to say I'd like it here. Nice people around who mind their own business. Then he asked what I paid for the school. "Whooee! Thirty-eight grand. You know what that old lady paid for it? Six thousand bucks!"

The first four months in the Bloomfield School we think of as our "commune" phase. We had given the Bloomfield villagers a few weeks to get used to longhairs in "their" schoolhouse before the rest of our entourage arrived. This included my mother, Elaine, on her annual visit, three apprentice filmmakers with films to finish, one resident actor, and Phoebe, our long-haired orange cat. No one's tenure would last more than a few months and we hoped local folk wouldn't feel threatened, though two of the apprentices even had dogs and it would take constant discipline to keep them from chasing neighbors' sheep or cattle.

Our real problems began in January when the heavy rains hit and our 1500 gallon septic tank, designed to handle four flush toilets, became inoperable, the electricity went out, and the leaning cypress, true to the dairyman's prophecy, came crashing down. An 80-foot cypress, standing, is impressive; supine it is awesome. It took out twenty feet of two fences and missed the vacant house by a scant four yards. On that first day, to show my intentions, I chopped off a dozen branches from the forest of foliage and dragged them over to our side of the fence. Thus began the ordeal of the tree. During the next six weeks I spent every spare moment axing and sawing limbs from this downed giant and hauling them over the line. Often I would come home from a full day professing filmmaking, change into my shit-shoveling, tree-chopping clothes, and be faced with new crises. The worst of these was the day Stormy told Elaine the town was going to burn us out.

Bloomfield was a large and thriving community in the late 1800s. The main street was once lined with a full mile of Victorian storefronts. In 1948 a fire consumed or ruined most of these buildings as well as the fire truck. Today only three of those buildings remain—Fireman's Hall, the Masonic Lodge and Stormy's Bloomfield Tavern, the only functioning commercial establishment in this erstwhile town.

My mother is a humanitarian, universalist, and free-thinker who likes to drink beer. Although she can't stand bars, Stormy's was the only place to buy a six-pack "to go" and she took to making a daily pilgrimage to Stormy's. On this particular day Stormy launched an abrupt interrogation. "What's going on up there? Who lives in that trailer? What's all that hammering at night? How many people live there?" Elaine calmly paid for her six-pack and walked out. Stormy followed her to the door and yelled one last thing: "You know, they burn out hippies around here!"

***

The problems of city folks like us adapting to rural life continued. The gods conspired against us, giving us the wettest winter in 75 years, which knocked out our water pump and left us with no pressure in the house lines. The pump house was two hundred feet away. At night, during a torrential rain, we determined the break occurred underground somewhere along that stretch.

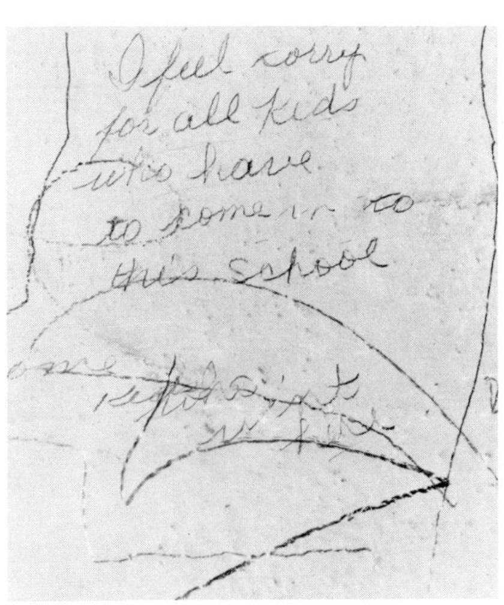

Closing day.

The VanMeters at home. "The common school is the greatest discovery ever made by man."—Horace Mann

Tatami mats help differentiate small living areas.

The next morning I went looking for free advice and luckily encountered "Big Earle" at a hardware store in the nearby town of Petaluma. He was the town's "electric expert" I was told, a big burly fellow with curly short hair, heavy glasses and a Pendleton shirt. His pockets held an arsenal of pens and paper, slide rule and a mini calculator. I guessed Earle was straight A's in science and played bass drum in his highschool band. After graduation, unable to afford college, he went straight to work and continued his studies through *Popular Mechanics, Scientific American* and the public library. He diagnosed and solved the conduit problem in a matter of minutes. He drew me a wiring diagram, wished me luck and even offered to drive out on his lunch hour if I ran into further trouble. Nice guy, Earle, though I didn't need him again: I dug a new 200-foot trench, laid the conduit as he instructed and, with a little wincing, threw the lever on the pump. It whirred immediately back into action. "Oh thank you, Mr. Wizard!"

Back to the tree. It took nearly two months to cut off all the limbs, using hand saws, axes and a small chain saw from Sears. Then came the major operation—the truncation of the trunk. No job for an amateur, even with large rented chain saws. When you attack a standing tree with a chain saw, it will fall. But as you cut through a fallen tree, it will settle and pinch the blade making it impossible to withdraw. What can you do but rent a second saw to extricate the first? I gave up on chain saws and spent $45 for an oldfashioned two-man saw. But that got stuck too. So I called Frank from "Frank's Apple Corner" who moved onto the scene with his son, jeep, flatbed truck, tractor, forklift and six chainsaws. Within three hours they had cut the mammoth trunk into stove-length sections and stacked them over on our property. For $75 I got two years of firewood that I could easily split with wedge and sledge.

As the last log was rolled over the line, my neighbor walked up through the calf lot. He hadn't spoken a word to me since the day we arrived. He had a hammer in his hand. "I'll fix the fence," he said. The next day, as we drove by, he looked up from feeding his mules and damned if he didn't flash us the peace sign.

All this time, all winter long, in weather foul or fair, I had been shoveling our wastes to rest, due to what I figured to be a flooded septic tank. It was spring before I could afford a $50 service call. A trip to the Building Inspector's office had shown me the septic tank's location—70 feet from the house on a line that ran under the driveway and basketball court. Farmer's Septic Service responded to my call with a pump truck and a strapping 19-year-old named Michael. He found the septic tank not at fault, but the line, through some years of disuse, plugged by dried solids. Together we rooted it out with steel rods. Believe me, there is no sound so dear to the heart of a retired school maintenance man as the "frooosh" of a happily flushing toilet.

There were other minor crises: the bus that balked on cold, wet mornings; Phil's dog killing two sheep belonging to a neighbor we had never seen until the morning he slid his pickup to a stop in front of the house and jumped out yelling, "I tracked that son-of-a-bitch back here and I'll give you fifty dollars to let me shoot him!"

Spring was around the corner and come hell, high water, sheep farmers, shitkickers, tree-choppers, building inspectors, mothers-in-law or mortgage-holders, I was planning my first real garden. My mother, the apprentices and the actor had left for warmer climes and Sandra was now pregnant with our third child. I sketched planting plots, ordered seed and found a local rancher to plow and disc a quarter-acre plot. One rainless day in May, with Lila and Benjamin carrying string and stakes, we laid out twelve one-hundred foot rows. Our other nextdoor neighbor leaned over the fence, introduced himself, and informed us the corner I had chosen for the garden had been a stable for the old school a hundred years ago. I knew, of course, it had

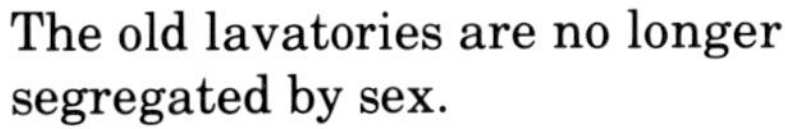
The old lavatories are no longer segregated by sex.

never been farmed, having been a schoolyard since 1860, and that manure tea from the cow pasture above had been seeping through the virgin topsoil for years and that, being the southeast corner, it got the longest daily dose of sunlight. But I hadn't expected hundred-year-old pony manure on top of all that.

The garden grew and grew as I watered, hoed and weeded. I roto-tilled, shredded and mulched; chased gophers and crows. Sandra started seedlings indoors—cabbage, broccoli, Brussels sprouts and cauliflower. She planted snow peas, spinach and carrots. By the end of August our secondhand freezer was full of corn, veggies and chicken. The shelves in the lunchroom were laden with 150 jars of goods, including "Dutch Spiced Red Cabbage."

After a very long and hard Labor Day, with a good doctor, a midwife, Sandra's mother and myself in attendance, Samuel

Ben editing film.

Desk parts discovered while rototilling.

A perfect place for neighborhood kids to gather.

Quanah Thomas VanMeter was born in the Bloomfield School. It was 4:30 in the morning and I rushed out to ring the bell.

As an aside, let me tell you about the bell. In 1860 the great-grandfather of the sheep farmer who wanted to shoot Phil's dog brought a bell around Cape Horn. It was installed in the Bloomfield Presbyterian Church which became the first school. (This church had its stable where our garden is.) When our building was erected in 1921 the bell was moved there, where it stayed until 1966. After Bloomfield School closed I have no idea what happened to the swings, teeter-totter and desks, but the history of the bell is well documented. Both ranchers and townspeople contended for ownership. The ranchers, organized as the Volunteer Fire Department, swiftly stole the bell. The townsfolk organized the Bloomfield Community Club and sued. Two years of litigation followed; suits and countersuits. The feeling of town thus persisted, even in the pastures, long after the town itself had all but disappeared. Finally, to resolve the controversy, a tower of telephone poles was constructed over a concrete wishing well, where the bell hangs today. Access is through a trap door with a large padlock. During the first year we lived here, I heard it ring but once, on New Year's eve.

On the morning Sam was born, I climbed to the roof of the bell tower. It was 5 a.m. and the dairy was already milking. I inched slowly up the slippery shingles until I could straddle the peak. Then I grasped the rope in both hands and tolled the bell for all I was worth. I announced to the sleeping dozens in the town and to the hundreds in the graveyard, "There's new life in the old schoolhouse today!"

—Ben VanMeter

OLEMA SCHOOL

The Olema School (A.D. 1915) was victimized by the first private owner who "modernized" it. He covered fine hardwood floors with linoleum tile and divided the large schoolroom into partitioned cubicles. Hannah and John rescued it from him, ripped out all his "improvements," and returned it to its original state after years of diligent restoration. The front steps are gently overgrown, children's voices echo again in the one big room. Morning sun streams through huge eastern windows, and a woodburning stove provides the only other heat.

Sarah, Naomi Sun and Maya Moon.

Hannah and her windows.

MOWRY'S OPERA HOUSE

As temporary City Hall during post-earthquake period.

Mowry's Opera House, constructed in 1879, survived the San Francisco earthquake and fire of 1906 and was used as a temporary city hall while the present one was built. At some point, what had been a three-story theater was sectioned off by adding a second and third floor. The loft, where Sandra and I lived in 1966-68, didn't officially exist. No building permit had been issued for it; it was built as a bareknuckle prizefight gym. Since all the police and firemen, city fathers and politicos were fight fans, the obvious site for the arena was upstairs in the temporary City Hall. It is rumored that here Gentleman Jim Corbett licked John L. Sullivan.

The loft was a unique open space—100 by 60 feet, uninterrupted by columns or pillars. Imagine 6000 square feet of floor space, with large windows on three sides and a huge oval skylight in the center where the fight ring had been, and where we put our bed. An attic above the loft enclosed foot-square timbers 60 feet long traversing the attic every eight feet, explaining how such a large ceiling could be hung without center supports. A friend floored in part of this space and lived here happily with the pigeons. This garret was larger than most houses.

We acquired the entire wardrobe of a defunct 50s burlesque house and often had our friends over to play dressup, roller skate, play with colored lights and watch films. We used many of the stage sets from the burlesque as

Six thousand square feet uninterrupted by columns or pillars.

moveable dividers to control the space. Our bed was a set piece fashioned like a large platter on which, undoubtedly, a damsel had been served up nightly for the delectation of the boys in the front row.

We rented this enormous space for $152 a month. Those were the days! We subrented a corner of the loft to Don Donahue who printed the first comics of Robert Crumb and Gilbert Shelton. And there, with all that space, I shot burlesque-nudie light-show epics, while Sandra poetically filmed the subtle changes of window light throughout a day.

The abundance of indoor space was inspiring. One night at Christmas time, while Sandra slept, I prowled the streets and picked up all the discarded Christmas trees I could find. She awoke, on our platter-bed under the skylight, surrounded by a forest of spruce, fir and pine!

Since we didn't own the building, we knew our days there were numbered, that redevelopment was inevitable and the building doomed. If we had confidence in staying, I would've painted the walls white, the ceiling black and covered the whole floor with tatami.

In cold weather we retreated into a small side room where we could hear the freight elevator with its ancient donkey engine, its cables creaking with the resonance of a cello. The building exuded a grandfatherly spirit we were sorry to leave.

MAX'S CASTLE

A three-story, turn-of-the-century brick building in downtown Pocatello is Max Richardson's empire. On street level are three store fronts housing his Mormon bookstore (formerly a bar with a mirrored back wall), a gift shop and Max's Unusual Books, the only place in town where you can buy *Screw* magazine.

Max and his family live upstairs in what used to be a flophouse hotel. His first intention was to renovate and reopen it as an upgraded hostelry. But government impediments were so discouraging to Max, whose patience with bureaucracies is nil, that he moved in himself. This ex-hotel resembles the set for a Bogart movie. The top floor is a mezzanine behind a polished bannister (I could almost hear shots ring out and see bodies crashing through to the lobby below). Max calls it his castle and restored it from shambles to shabby elegance for less than a thousand dollars. He shares it with his wife, three of his seven children, and two employees. He conducted us through a cheerful dining room and up to the second floor lobby, now his living room. As the photos demonstrate, living in an art deco hotel lobby has not changed Max into a lounge lizard.

Max's Pocatello empire—two bookshops and the Marion Hotel.

Marion's lounge

Marion's lobby

We told him about Sunshine Camp. "What we have here is similar," he said, "but where you have a separate house for everything, plus all your outside space, we have it all under one roof. This is our yard, our garden, our workshop. If we want to store bicycles, we just open up another room. Same if you get tired of the room you're living in—just open up another one." The thick walls make the old building a quiet island in the noisy storm of downtown.

Max in his castle.

RENO HOTEL

Reno Hotel, rear.

A gloomy interior.

Old doors converted into a sleeping platform.

Two filmmakers, Scott Bartlett and Tom DeWitt, leased the basement of the Reno Hotel in the late 60s. The former kitchens became the site of experimental filmmaking—shooting, editing, screenings and meetings. Ken Kesey and his Merry Pranksters held a Halloween party there while Kesey, to satisfy a condition of a pot possession charge, publicly denounced the evil weed. Bartlett and DeWitt are long gone from the Reno, which has since housed a succession of artists. Today, two young women are holed up in the gloomy interior; they punched through the ceiling to provide access to a couple hundred rooms upstairs, though most of the windows are gone and glass litters the hall.

The building is a dead ruin and the tenants are affected by the sense of inhabiting a corpse. Rescue is not always possible or even desirable. When something is dead for sure, it should be buried.

LIGHTHOUSE

Perched on a bluff high above the Pacific, this brave little frame lighthouse was a Coast Guard installation until the early 40s, then was sold to private owners, who constructed a comfortable small house around it. Its occupant guided us through, opened an inside door, and revealed the captive lighthouse, a door to the interior, and a steep ladder-stairway. We climbed to an eyrie/bedroom which contained the treasures of a young man in his late teens and at least ten square feet of floor space. Imagine waking in the night to howling wind and rain to watch window reflections distort as the gale bows the panes in and out, with an occasional quick dusting of spray from the breakers far below.

Nowadays, romantics who set their sights on lighthouse-keeping must reconcile themselves to being unrequited—the days of the manned lighthouse are nearly at an end. Most have now been automated and need little maintenance . . . but keep a weather eye for ones that no longer light up at night. You may find one in need of rescuing.

SILO

Rural highrise in upstate New York: a unique opportunity for vertical living. Though still under transformation, this five-story grain silo shows careful planning and assiduous attention by the builder.

Warm air flows up from the kitchen on the bottom floor, through the living room, guest room and study to the bedroom on top, taking full advantage of the chimney effect for heat.

The rescuer in his fourth floor study.

ROOST

Petaluma, California, was once the "egg capital of the world." Eggs were laid by the thousands on chicken ranches like this one in the benign northern California climate . . . a reasonable number of eggs with yolks the color of oranges.

Then science automated the chicken, confined her to battery cages off the ground, fed her chemical stimulants and crowded her into gigantic concentration camps. Man's ancient companion, the loyal chicken, now produces more eggs than ever before, but with yolks the color of pale urine.

Thus, chicken ranches came on the market. Myron and Donna bought this five acre one which included a main house, a processing and candling barn, and ten 20 x 100 foot redwood chicken houses with good roofs and floors. The soil, after years of chickens, is incredibly fertile. Myron dismantled several of the chicken houses and used the lumber to handsomely panel the interior of his living quarters. One of the long buildings houses his vast storehouse of hardware and junk gleaned from the county dump. Other buildings provide space for any of a thousand future projects.

"Someday, my son, all this will be yours."

SPRING HILL

Newell and Ruth Hart's dream of being chicken farmers was thwarted when their horde of baby chicks, brooded in the attic, succumbed to various forms of chicken sickness, poultry crowding, and marauding rats. That left them with a newly completed chicken house, a mansion by farmyard standards, and the dilemma of what to do with it. So Newell enlisted the services of Lott Vienwig, southern Idaho's most eminent housemover, dragged it to the top of Spring Hill and transformed it into a nightclub, a tribute to his skills with wood and stone. It also fulfilled his love of evening beer accompanied by traditional jazz and jam sessions. But the rigors of life for a barkeep in rural Idaho where on-sale liquor is illegal forced them to lease out the club and move away. Two decades later, they returned to Preston, Idaho, repossessed the club and have spent recent years shaping it to their hearts' desire. Newell also edits and publishes the *Cache Valley Newsletter* and has become the region's historian and archivist. He struggles to save historic buildings from condemnation, such as an old school building he is restoring almost single-handedly. From the ruins of those demolished he scavenges interesting bits and pieces to fit into his place on Spring Hill. With the joy only a confirmed recycler can know, Newell told us, "After all these years of work, the county assessors came out to inspect—and lowered our taxes by a third."

Ruth in the pantry.

Newell at work on chimney.

Door made of old desk tops rescued from another project—a schoolhouse.

The new chicken coop.

BODEGA CREAMERY

The original building was erected in 1895 and, over the years, grew like Topsy. Closed as a creamery due to the Civil Defense brown-out during WW II, it has been used, in more recent years, as storage space for boats and campers. With a county moratorium on building due to a water shortage, it had become a realtor's white elephant.

Then two artists, Bill and Wopo, took over. It could be ideal for them, but wasn't in the beginning. Creameries are built to keep milk cold. Their huge spaces are hard to heat. They have concrete floors and walk-in refrigerators. Bill and Wopo moved into the refrigerator. As the only insulated room, it was the easiest to heat. Skylights were broken. Debris littered the floors. One room was so full of junk it completely hid the door to another room. Walls had been built around the chassis of a derelict truck that Bill had to cut to pieces with a torch to get out. Wopo, a lithographer, didn't like it there. She wanted a real house. During their first winter in the creamery they huddled around the fire and had constant arguments. "You'd think an artist in a huge space would be inspired to do gigantic pieces," Wopo said. "I sat next to the fire and did the smallest drawings I had done in years." Bill had to rewire the place completely. He tried to keep the exterior inconspicuous, even antiqued new water pipes which could be seen from the road.

Today, a year's work later, Bill and Wopo each have their own sunny studios. Wopo also has a lithography print shop large enough to hold classes. She adapted to the cold. "A creamery needs cool space; so does lithography. You have to sponge your stone, keep it damp. If the atmosphere is hot and dry, the stone dries out too quickly." Now they have galleries, a film-slide projection room, and an indoor workshop large enough to build a Sherman tank in. A Japanese bath is being built from "free and scrounged materials." Much of the concrete floor has been covered with wooden flooring, or with a mosaic of irregularly shaped pieces of gunstock walnut. Work has begun on a "New York type" apartment in the attic. One of these days Bill is going to paint back the faded letters of the ancient sign on the side of the building: "HOME OF BODEGA'S BEST."

An Alice-in-Wonderland view into the cooler/dining room.

Wopo at the litho press.

VENICE IN AMERICA

Here is the phenomenon of almost a whole town being converted from public storefronts to homes and live-in studios. Venice, California, once a fairly typical beach community with shoe shops, garages, stores and arcades, is being vigorously recycled by a new breed of artists and romantics living behind boarded-up show windows. This is logical to those who know the history of Venice, studded with transformations wrought by recyclers and land-grabbers of other stripes.

In Southern California, 1904 was much longer ago than it was in most places. In that year, on a deserted stretch of beach 25 miles from Los Angeles, a wealthy visionary named Abbot Kinney trod the pristine sands and was suddenly struck by the inspiration to create a "Venice in America." He was determined to build a replica of the Italian city of the Doges, a great center of education and culture, to provide, in his words, "entertainment for the cultured, good music for the masses, wholesome playgrounds for children and artistic homesites for writers, musicians, sculptors, painters and retired capitalists, along the romantic winding canals."

He wasted no time and bought 160 acres of sand dunes and marshes and hired a team of architects and engineers. They constructed an elaborate system of canals filled with sea water, which flowed in at high tide and was retained by an ingenious system of locks. Along the canals, Kinney's minions erected scores of Italo-Californian stuccoed bungalows and a clutch of quasi-Renaissance public buildings. His master stroke was to import a fleet of genuine Venetian gondolas, complete with two dozen gondoliers. He created this whole scene by July 4, 1905—from vision to fruition in just one year. Disney would have loved it.

But few others did. Neither the cultured nor the masses had much interest in Venetian living, and the artists were too poor. Even his eleventh-hour importation of Sarah Bernhardt didn't draw flies. Meanwhile, an incredible beach-front land boom was occurring everywhere else along the coast. As Kinney threw good money after bad in an effort to revive his dying turkey, Huntington Hartford was raking in a fortune a few miles south at Redondo Beach.

As cultured as he was, Kinney was first and foremost an opportunist. If folks didn't want the finer things in life, he reasoned, then by God, we'll give them exactly what they *do* want. So, in 1906 he converted Venice in America into an amusement park, complete with dance halls, gambling dens, roller coasters, freakshows—mostly imported from the defunct Portland Fair. It was a fantastic success. Homesites sold at last, and southlanders built tacky cottages along the canals, where transplanted farmboys in homemade bumboats taxied tourists to the hot dog stands.

"We committed suicide," commented Thomas H. Thurlow, last mayor of Venice, after it was annexed by Los Angeles in 1925. The first move by L.A. authorities was to condemn the dirty canals and fill them in. Three hundred residents, armed with picks and shovels, couldn't stop them. The arched bridges were torn down and boat landings replaced by gas stations. Then a small oilfield was discovered in the midst of all this and the dreamland fun-house was replaced by the gooey reality of oil derricks and petroleum miasma. Finally, the beach was quarantined because of oil sewage.

Once a fairly typical beach community with shoe shops.

Then came the 50s when the oil played out. Venice, much the worse for wear, experienced a hiatus now its beach-front property was worth its weight in gold, rather than oil.

Venice today bears many of the happy earmarks of the ghetto (lively streets, neighborhood identity and rapport, community projects such as free legal and medical services, work collectives, art and theater projects) as well as the gloomy ones (poorly maintained dwellings, overcrowding, and the undying enmity of the well-to-do). Moreover, the ghetto of Venice is built on valuable real estate. The population is spread out into single-family dwellings, some multiples and a flourishing street scene. The inhabitants are racial minorities, the poor, artists and hippies. Their domiciles range from elegant to sub-standard, but all are relatively inexpensive. They have a sense of community, in spite of a certain incidence of petty crime, and show the possibility of comfortable low-income existence. The scene is ripe for exploitation. The fate of the Warsaw Jewish ghetto in WW II, where they crowded in the riffraff, closed the gates and ignored them all, was the epitome of solving ghetto problems. One modern American solution is through strenuous urban redevelopment, in which authorities declare an old neighborhood a disaster area, force out its inhabitants, tear down the buildings to creat vast open areas of ruins that function as short-lived crime parks before the plastic high-rises shoot up.

Another version of the scenario is being prepared for Venice. It goes something like this. Taxes and rents go up. The poorest citizens are forced out first. Newcomers, less involved in the community, take their places. Property improvements become clandestine as no one wants to be discovered upgrading his holdings and having taxes rise still further. Small businesses go under, diminishing the street life. Real estate speculators make offers hard to refuse. They upgrade old buildings with cheap cosmetic facelifts and rent them at inflated prices. Other old buildings are torn down to make way for high rises and condos. Through this process, the "old neighborhood" is converted into an urban horror—faceless building-block buildings, deserted streets, a million windows staring blankly at the Pacific.

We went to Venice recently to see the sights. I found it to be amazingly diverse, reminiscent of North Beach in the 50s or Haight-Ashbury in the 60s, but with more sense of the outdoors. Vestiges of earlier times still exist, such as numerous court streets closed to vehicular traffic. It was a Monday and everyone seemed so busy—exercising, running, walking, bicycling, playing paddle tennis, roller skating, weight-lifting—all tanned, right out in the open—a real public celebration.

We spoke to Irina in the Environmental Communications office, and she described the increasing difficulties of living in Venice. "Rents are rising, condos are threatening," she said. "Resistance from the populace is weakening because so much of the original ghetto population has been replaced by reclusive artists. They don't seem to care what's happening in the community. They hole up in old commercial locations, boarded up behind obsolete storefronts."

I wondered why. Irina explained the value of undifferentiated space, room where artists can construct visions from within

A new breed of romantics living behind boarded-up show windows.

themselves. She spoke of places with toilets completely enclosed in the middle of a large room. She seemed resigned to the inevitable swallowing of Venice by Los Angeles.

We strolled along the beach front, in late afternoon shadows. The air smelled good, and the smog hung a half-mile to the east, held there by offshore breezes. The ghetto by the sea. We walked through this mixture. Someone smoked a joint. I heard a guitar. When Venice goes, the shore will be owned by the rich and the government.

Small businesses go under, diminishing street life.

We strolled
along the beach
front in late
afternoon shadows.

Someone smoked
a joint. I heard
a guitar.

VENICE

This is Frederick Eversley in the doorway of his rescued laundromat. With his patience he could've been a great telescope lens grinder, but became a superb polyester sculptor instead. Inside this mean brick facade is a sparkling white gallery/workshop. The polishing room (not shown) is nearly hermetically sealed. His luminous pieces go through 14 abrasives to finish, necessitating a meticulous purging of each previous grit, carried smartly away through the old laundry drains. He lives where he works and works where he lives. When he sleeps it's up that ladder behind where he sits. Three thousand square feet of laundromat converted by his own hands.

Next door to Eversley is another breed of artist, Gene Sturman. His fridge looks diminutive in the old meat locker of Irv's Family Market. The antique stamped sheet metal ceiling was revealed when he ripped out a celotex overlay. "It was absolutely fortuitous," he said, "being a sheet metal sculptor myself." Basically, the work involved stripping to the bare walls, hauling old shelving to the Goodwill and cleaning up "all the insidious meat tenderizer lying around."

The dilapidated facade of another era—the Ace Garage. This was rescued by Don Griffin, silk screen printer for artists. The interior was a grimy mess after decades of fumes, grease and oil. He put on a mask and firefighter's suit and sandblasted everything in sight, including the old grease pit, now a gleaming hot tub between dining table and kitchen. Lots of studio room for compatible people—serigraph area and space for dyeing fabrics and printing batiks.

The old Venice firehouse—gallery downstairs and home upstairs where firemen once played cribbage and waited. The fire pole? "The firemen were smart enough to take it with them," said Ian, who transformed the barracks into a most comfortable, soundproof home where the alarm never rings.

TRAINS AND SUCH

As trains continue to join Conestoga wagons and chariots in museums of outmoded vehicles, there'll be more opportunities to turn these wheeled behemoths into comfortable living spaces. Some, of course, were carefully built and furbished for that very purpose—recall the extravagant private railroad cars of last century, and the enduring caboose. Yet even a freight car has charm, and offers a spacious and well-insulated compartment, solidly built and high-roofed enough for sleeping lofts.

The logistics of securing one's very own railroad car are not always clear-cut and predictable. Some railroad companies sell old cars to the highest bidder, others sell them in large lots to scrap metal companies. If you can buy a car from your nearest railroad company, you may find a real bargain—they often charge as little as $1500. Salvage companies charge more—$3000-5000 for cabooses, boxcars and reefer cars, $5000-7500 for passenger and baggage cars.

Having purchased a car, you are faced with the problem of transporting it to its new home. If your car will still roll on tracks, it can make at least part of its journey as part of a train, at a cost of several hundred dollars to hook up, plus around a dollar a mile for the trip. But most old train cars are no longer trackworthy and must be transported entirely by truck. In some cases, this will mean *three* trucks—one for the car, one for the wheels, and one for the crane needed to take the car off the trucks and set it in place. One rule of thumb I've heard is thirty dollars per vehicle per hour, portal-to-portal.

Problems of transport become compounded relative to both height and length of the car being transported. Trucking companies must figure a route to your site that has no bridges too low for your car to pass under; ease of access at the delivery end has bearing, too—just how long will it take a truck driver and crane operator to ease the car onto its ideal location?

It may take a little wheeling and dealing, but the privilege of owning a piece of American history should more than offset the inconvenience.

TRAIN STATION

With its loading dock and baggage trucks filled, this New York station could easily be mistaken for your point of departure. Actually it is the home of a photographer who loves trains.

Freight and shipping area provide a roomy studio and upstairs sleeping loft.

Instead of times and tickets this room now serves dinner and drinks. Salads are passed through the ticket window, behind refrigerator.

PARLOR CAR

End of the last run. This stately passenger car once made the run from San Francisco to Jenner-by-the-Sea. It was retired early because of a congenital sag in its midsection, and so escaped the ravages of long service.

This car and those on the next page were originally purchased by a railroad buff and general collector of oddities, and at present provide low-rent housing to their happy occupants.

CABOOSE COUPLET

Twin cabooses coupled together make a single residence: kitchen and dining area in one, living room and bedroom in the other. Built in 1904, they became the property of a physician in Southern California who attempted their restoration. They were brought north by rail and truck a few years ago.

Long, narrow vistas give a feeling of expanse in a small area; simple decorative inspiration and a touch of threadbare elegance keep them happily tenanted.

This tongue-in-groove craftsmanship would be beyond budget for most home builders today.

TROLLEY

This streetcar, poking its nose out of a flowerbed in a small California town, is also a rental unit. Its rear end, hidden amid bushes, is constructed from odd bits and pieces of two railroad sleepers, grafted on. The day may soon be upon us when the charm of Victorian transportation is preserved only in residences like this.

CABOOSE

Wheel carriages removed, this rolling caboose has found peace in Death Valley.

MITCHELL ROAD STABLE

An old stable, formerly belonging to a gentleman farmer, now provides a grand home for a Rochester artist. This century-old building has been meticulously preserved on the outside. Each room inside has been assigned a new function—the old tackroom is now a kitchen, the hayloft a dance floor, and the feedbin a cloakroom.

The small thresher on the left is one of several antique farm implements now used for decorative purposes.

Carriages and wagons once filled this room. Now, half living room and half dining room, fine foods and conversation occupy the space.

A look down the breezeway shows each stall refinished and providing such services as bathroom, bedroom, library and sewing room.

The bathroom retains the horses' watering trough, now a bathtub.

SAN FRANCISCO CHURCH

Small churches, like old schoolhouses, seem to make the transition from public to private use quite naturally. Shown here are three metamorphoses—urban, suburban, and rural. This neighborhood church nestles comfortably among other Victorian homes in San Francisco.

The liturgical quality of the interior lingers only in the elaborate truss work and leaded windows.

SENECA FALLS CHURCH

With its cross removed, this suburban church is indistinguishable from other houses on the street.

Inside, the original use is apparent, though a kitchen has replaced the chancel and the choir loft has become a sleeping place.

SILVER CITY CHURCH

Once a house of worship for miners and their families in Silver City, Nevada. The congregation disappeared when the rich ore gave out. Now it's home for rock singer Lynne Hughes and media personality Travis T. Hipp.

Home in the belfry.

DRUIDS HALL

Another good bet in recycled institutions is the fraternal organization meeting hall. Many old lodges are still flourishing, but others have moved on to more modern quarters or have been unable to compete with prime-time TV. Their highest popularity came at the turn of the century, and even small communities often boasted elegant Masonic, Elks, Odd Fellows or Knights of Pythias halls. Today, these small-town facilities may be seldom used or abandoned. Sometimes there's enough public interest to convert these wondrous white elephants into community centers—but sometimes not, as was the case with this fine old Druid's hall. It was converted into a spacious family home with no structural changes at all. The owners are proud of the dais which still stands in the meeting room, where once stood the ornate throne of boss Druid. Now, it holds the residents' most elegant chair.

Rescuers Bob, Victoria, Omar and Tasha.

Latter-day "druids."

The noble Sandcastle Gallery was a Foresters' hall until the 70s—now it's a colorful arts-and-crafts gallery downstairs with living quarters on the second floor.

SANDCASTLE GALLERY

Lodge halls, by the way, often fulfill the twin desires for generous space and the need for private and cozy living areas, offering a large main hall as well as small inner initiates' chambers. Check your yellow pages for religious and fraternal organizations, scout them out and make discreet inquiries if the buildings have a feeling of disuse.

OAK BARREL TRIPTYCH

Barrel living provides a perfect use for those vats too large to be snapped up by hot tub builders. A vessel built to hold thousands of gallons of liquid is built to last. To people who live in tipis, hogans and barrels, the circle is a sacred shape.

This unique rustic home is constructed from three enormous wine settling vats, the trefoil shape augmented by connecting closets and entry hall.

THE WATER BARREL

Another variant on container living, this water barrel was purchased from the Southern Pacific Railroad at a bargain price. It was installed on a small ocean view lot and converted into living quarters at minimal expense. This simple structure is actually big enough to house two couples.

PALOMAR SKATING RINK

This amazing structure began life in the 30s as a resort-area dance hall. Big bands were big business, and during its early years, famous bandmasters Tommy Dorsey and Glenn Miller played the Palomar. Interest in swing dancing declined, and the Palomar was converted into a roller skating rink. Interest in roller skating declined, and the Palomar closed its doors a few years back, remaining vacant until Carrie McClelland and her son Scott, searching for a house large enough to accommodate both their lives, suffered an auto breakdown nearby.

Scott went in search of mechanical assistance, but came running excitedly back a few minutes later, shouting "I found it! I found it! It's a roller skating rink!"

The McClellands are still trying to figure out exactly what to do with the Palomar, with its walk-in fireplace, hundreds of rusty rollerskates, and six thousand square feet of maple flooring. They hope to build lofts for bedrooms and leave much of the vast floor space uncluttered. Scott is an enthusiastic amateur athlete, and wants to set up a gymnasium. He'll never lack space to ride his unicycle on rainy days.

An idyllic setting.

Complete with hundreds of rusty rollerskates.

SERVICE STATION

Once it was called the White Palace—a service station/garage/restaurant along the old Redwood Highway in Northern California. Now Larry Grasso lives in the gas station, enjoying his low-rent bachelor digs, and was tickled by our interest in them.

Before freeways and interstates, roads all over America were studded with these Mom-and-Pop enterprises. With the advent of turnpikes and multi-national corporations, most of them have been abandoned. Check your by-passed local highways when shopping for defunct institutions.

CHEMICAL 5
S.F.F.D.
1757

CHEMICAL #5

Chemical #5, in the Haight-Ashbury district, was rescued ten years ago by a filmmaker who lived upstairs in the firemen's dormitory. Since then it has been preserved by a succession of filmmakers. The free-standing fireplace with its long stovepipe is reminiscent of the firepole which had to be removed, ironically, to satisfy fire code regulations.

Most of San Francisco's old firehouses have been rescued from the indignity of the wrecker's ball. Childhood dreams of being a fireman, sliding down a pole and rushing to the rescue aboard a red hook and ladder have probably inspired grownups among us to collect firehouse memorabilia, including firehouses themselves.

The firetruck garage downstairs is now a spacious, sunny apartment. People living in rescued spaces usually hold warm fondness for their buildings' history, but none more than firehouse dwellers. The fireman has always been everyone's favorite public servant. In the course of our research, we never encountered a rescued jail.

LAST RESORTS

A generation ago, many aspects of American life were much simpler than they are today. Take summer vacations. Families began planning a short time ahead. Dad liked the seashore, Mom favored the mountains, and teen-aged kids wanted to be where the action was. On the appointed day, after much compromising, the family loaded the car and drove off to a *Resort*. Often, it wasn't much farther than where city stopped and "country" began. A typical resort would feature a number of fair weather cabins, falling somewhere between funky and fancy, usually called "rustic." There would be a restaurant, snack bar, dining hall or grocery store, a central meeting hall, plus amenities like a swimming pool, sports areas, tennis courts, horse stable, lake, adjacent beach and hiking trails. Evening pleasures were offered—games, dancing, getting drunk, shuffleboard, community singing, marshmallow roasts. A resort was a place to spend a lot of time outdoors and loaf. Here, people got sunburn, poison oak, didn't shave every day, wore comfortable old clothes, made new friends, and let their hair down. That's what vacations were all about, making you relaxed and renewed, ready to put your shoulder to the wheel for another year.

All that's changed. As populations grew, people began feeling more crowded and mass media took over. Nowadays, when vacation time comes, one goes to Las Vegas or Florida, takes an economy tour abroad, cruises the U.S.A. in a Winnebago, backpacks in the Sierras with a thousand dollars worth of camping gear, or simply stays home to watch reruns.

Consequently, despite the encroachments of suburban housing on country property, many of these old resorts are still around. Sunshine Camp, for example, was put on the market because its earning potential didn't meet its costs, and the dedicated people who gave it essential vitality and interest had died or given up. Some of these resorts, considered white elephants by their owners, hit the real estate market at bargain prices. Who, after all, wants to buy an unsuccessful business? Someone who's not going to use it in the same old unsuccessful way, that's who.

Resort properties are usually run down and have to be put back in shape again. Some have been long deserted and suffered the depredations of time; some have been vandalized. Yet for those who can make use of such space, they offer great advantages. All the pleasures people once enjoyed are still there—swimming pools and stables, play areas, trees and trails, and the buildings. But many were designed for summer use only and must be prepared for winter with heaters, insulation and weather-tight windows. Modifying to all-year use is usually the major effort in adapting a resort to residential use. So, if you're looking at a charming hotel on a bright summer day, with trees full of birds and squirrels, and the realtor is beginning to rub his hands and smile a lot, temper your enthusiasm with careful appraisal of how it would fare in winter.

Though both are picturesque, there's a basic discrepancy between a charming old country cabin and a total ruin. Make sure repair is possible and within your means.

Resorts usually include a number of living spaces, making it possible to recoup some of your expenses by renting out buildings or rooms. You may need to add a few more bathrooms and kitchens to qualify yourself for the role of a rural landlord. Be non-threatening, if not downright friendly, pick your tenants with care, and assess the attitudes of county authorities toward your proposed life style. Be sure you're not flagrantly violating anyone's rules—building inspectors', public health folks', the police's. Sonoma County, for example, had what its authorities considered a bad experience with hippie communes in the 60s. This doesn't mean you can't *have* a commune in Sonoma County, only you should try not to *look* like one.

You must hurry. If resort lifestyle appeals to you, start looking now, while there's still time. Resorts are growing in popularity once again, with different emphases this time. This growth is caused by interest in what's sometimes known as the Human Potential Movement. Spiritual organizations, gurus and followers, meditators, massagers, yogis, deep breathers, Gurdjieffians, estians, Jesus lovers, gestalts, primalers, sunbathers, sunworshippers, nudists, sexmonauts, druggers, astral travellers, UFO watchers, food faddists, musicians, dancers, theater people, writers, chanters and glossolalia experts—they fill the woods. Virtually every one of them wants a place—a center, community, ashram—somewhere they can all get it on together. We bought Sunshine Camp two days after it was advertised in a local newspaper. The realtor told us lively interest in it continued after we concluded our deal and that it mainly came from so-called growth centers looking for headquarters. So don't delay. Watch the papers. Talk to realtors, especially those who indicate any specialization in unusual or commercial properties. Contact the chamber of commerce. Tell them you're interested in resort and hotel properties, but *not* that you intend to convert one into residential quarters—they hate to see commercial businesses pass from public use. Check the public library for past or present telephone directories under "Hotels," "Resorts," "Parks." An old phone book or city directory can lead you to old resorts you may have never known, even in areas that are familiar. Churches and fraternal organizations often run resorts for their members and sometimes are open to inquiries. Familiarize yourself with areas around parks and other recreational land; you'll sometimes find past resort connections.

Some old resorts are being brought back to their original use again—I know of two nearby. They offer pleasant natural accommodations, good food, quiet relaxation and reasonable prices. So find your resort before it sprouts a sign reading "Under New Management."

CAMP IMELDA

Greg Crisp, a builder, stumbled upon Camp Imelda at just the right time. It had been a Roman Catholic girls' camp for years, lovingly maintained by a craftsman/caretaker. But the church wanted to withdraw from the summer camp

business, and put Camp Imelda on the market with a six-figure price tag, minus all movables, furniture and fixtures. Greg offered less than half the asking price, and thereby bought the camp.

What do you do with two-and-a-half acres of summer camp under giant redwoods, including more than fifty buildings, several fully equipped residences, a 250-seat meeting hall with elevated stage, projection room and movie projector, a library of several thousand children's books, a kitchen far better appointed than most restaurants', a grand dining hall, a chapel with fully draped walls exuding a fine air of sanctity—what do you do with such a rustic empire, if you have no interest in teaching little girls to swim in the nearby river or cooking them hot cereal every morning? Greg is asking himself the same question and, until he figures it out, he and his friends are simply enjoying life in their woodland village.

In these times, with the so-called Human Potential Movement the hottest thing to come down the pike since Charles Atlas and Dynamic Tension, Greg has already been able to rent facilities on an occasional basis to self-help groups, and hopes ultimately to establish some kind of growth center at Camp Imelda. Here, certainly, is a recycled institution big enough to contain even the most ambitious projects.

Even a chapel.

And a pristine, beautifully equipped kitchen.

SUNSHINE CAMP

The Happy Village

At Sunshine Camp, with a normal population of 12, the village parallel is evident. The inhabitants rarely get together until evening. Everyone, intent on a task, is isolated during daylight hours. I wave to Molly as we move through the bright quadrangle, I call to George as his welding torch flares under the carport—I move on and they are gone, sometimes for hours. Yet I hear them . . . indistinct chidren's voices repeating something over and over, a saw screaming, a stereo playing. Then evening comes and members of each nuclear family talk about food—who will or won't be eating, who will prepare each of the meal's various components. We gather for supper and share the day's experiences, then each of us drifts off to his hut as night falls. In darkness I walk across the quadrangle, trip over a toy, see lights in windows, unfamiliar cars here and there, hear voices and music. Needless to say, we're happy here.

Think of this as a village—an extended family village composed of my wife's brother and his wife, my wife's sister and her husband and two kids, besides ourselves and our three youngsters, an apprentice or two, and sometimes a friend in need.

When my wife, Alexandra, first proposed that all of us get together, pool our resources and buy a piece of communal real estate, I did a lot of hard thinking. It could make sense, I decided, if by splitting the mortgage three ways we could have something better than we could afford separately. I'd known my prospective commune mates for years. Alexandra's brother, Kelly, was still wet behind the ears when I first met him and I had watched him develop into the musician/photographer/artist/filmmaker/inventor-of-a revolutionary-process-for-animated-movies that he is today. His wife Rosana—librarian, potter, and delver into the occult—I'd known her a while too. And of course I'd known Alexandra's sister Molly and her husband George since the days before they built a boat and sailed it across the ocean, then sailed it back after a few years and took up teaching *t'ai chi chuan.* If I couldn't get along with these people, it was obvious I couldn't get along with *anyone*.

There were months of appointments with realtors, want ad orgies, frantic phone calls, near misses, and weekend property-inspection outings. Land values in our area of search were rising fast. In general we found ourselves looking at low-profile agricultural properties, rarely more than one residence, a few outbuildings and some acreage. Mom-and-Pop subsistence farms, really, which would require construction of new dwellings we couldn't afford. Then Molly stumbled on Sunshine Camp and the whole extended family descended upon it with high hopes.

Sunshine Camp was founded in the middle 30s by an enterprising Roman Catholic priest, and bustled along for three decades until he died. Local well-wishers tried to keep it open, but were defeated by the IRS who refused to grant them non-profit tax-exempt status. (It used to be relatively easy to dream up a front to obtain

The boys once gathered every morning for flag salute.
Sunshine Camp kids today.

tax-exempt status to satisfy the government, but it's not easy anymore. So those who plan to take over a defunct non-profit institution and enjoy its exemptions had best make thorough investigations and read all the fine print before signing anything irrevocable.) Thus, the property came up for sale. It was split into two parcels. The smaller one, a 4½ acre piece including most of the buildings and two acres of apple orchard, was traded from one owner to another for several years, during which time the fourteen buildings, rapidly deteriorating, were rented out at nominal fees. It had become a pleasant, low-keyed enclave, a commune/motel crossbreed. When we arrived on the scene, the camp sheltered more than 20 full time residents.

When we first drove through the front gate all my village-dweller genes gave a glad little hop. I was simply conscious of a broad, open area combining bright sunlight and dappled shade, and one house after another, all rather homey and decrepit. And that over there, could it be a *chapel*? Yes, indeed, with a genuine stained glass window and confessionals. We were all enthralled—but not the people who lived there. When we explained we were in the process of buying the place they reacted with surprise and consternation. We realized to our dismay the present owners hadn't even told the tenants the camp was for sale.

We retreated to the realtor's office, signed papers and wrote a sickeningly large check. That began an entire year of legal divagations before Sunshine Camp (and its several mortgages) became ours, a chronicle of paperwork and manuvering of interest only to a dedicated student of bureaucracy. We learned that transfer of ownership in the case of ex-institutional properties is likely to be far more complex than buying a tract home and dealing for an FHA loan.

Finally, it became ours—a long four acres, the front half level and dominated by a rough quadrangle of old buildings—the original century-old farmhouse, a ramshackle kitchen/mess hall, the chapel, and a cluster of bunkhouses, cabins and utility buildings. The rear half of the land rises steeply to the crest of a hill and contains 80 elderly Gravenstein apple trees. There is ample room for everyone to live and work: a weaving studio for Alexandra and her 13 by 20-foot high-warp loom; a pottery workshop for Rosana; a film animation studio for Kelly; a wood and metal shop for George; room for Molly to work on leaded glass; and a bicycle workshop for my son Hobart. The chapel has become the space for Molly and George to teach *t'ai chi*. Besides all this, we have separate little houses to live and sleep in and all the bathroom and kitchen facilities we will ever need. Nothing like an institutional property to provide an abundance of *facilities*.

It cost us some real pain to dispossess the tenants. They were paying moderate prices for very pleasant accommodations and some had been there as long as three years. We knew it was actually the realtor's responsibility to do the evicting, but decided to do it ourselves as gracefully as possible to minimize trashing and pilfering by the evictees, which can be a real hazard to the new owner of institutional property. When tenants informed us various items were their personal possessions which they were willing to sell or trade for unpaid rent, we accepted most of the time, and it turned out to be worth it. We ended up with a largesse of stoves, refrigerators and miscellaneous useful items. We exchanged smiles and best wishes with the last tenants who moved out some nine months after we took possession.

As we familiarized ourselves with the camp we discovered other ex-institutional benefits. The giant old mess hall stove, for example—dreadfully dirty, but where

Sunshine Camp in an earlier era.

Today's camp kids play kickball.

everyone goes to cook huge quantities of messy stuff such as apple juice from the orchard (we pressed 250 gallons this season), buckets of tar for winter roofing, and pots of dye for Alexandra's miles of looped yarn. We found a walk-in cooler, inside which lives an old grocery store ice cream freezer that keeps meat well frozen and leaks enough cold to keep the cooler at an ideal fermentation temperature for George's home brew. When we moved in the camp shower house contained its own water heater with four stalls enclosed to shoulder height. For our first Christmas together, George converted part of the building into a 5 by 7 foot concrete hot tub, thanks to Leon Elder's inspiration. Believe me, if you plan to live with your in-laws, there's nothing wiser than a hot tub where everyone can get together, be naked, and soak away frets and tensions in good hot water.

There were, of course, many

Kelly's portal symbolizes the coming together of three families.

disadvantages: lots of roofs to spring leaks, septic tank disasters that'd chill your blood, hundreds of electrical connections to short out, pipes to rust, broken windows to replace. So realize you can have anything you want, if you're willing to pay the price, one way or another.

Let me interject a few general thoughts on institutions. They always seem to retain a flavor of their original use. I'm not sure an old jail or madhouse could ever be made very pleasant to live in. Once, in San Francisco, some friends lived in a building which had been an orphanage. It had been divided into relatively spacious and comfortable apartments, the neighborhood was good and the rent cheap. But a pervasive sadness hung over the place and I was never fully at ease while visiting there.

The coming of new life is a sure-fire catalyst for the tradition-making energies of a community. Shortly before Rusty was born, I saw Molly standing nude in the doorway of her house (once the camp infirmary), her belly gigantic but still extraordinarily graceful, taking her body through the ritual movements of *t'ai chi* in preparation for the coming effort. I knew all Sunshine Campers were tuned to this event, and I thought, "That kid isn't even born yet, and it already has a hell of a lot going for it." Then, in the night, we all crowded into the tiny building marked "storehouse," which is George and Molly's bedroom, and watched George, a student-nurse friend, and a midwife help Molly toward the moment of birth. And finally, amid laughter, tears and goggle-eyed kids, little Rusty came out to take his best shot.

Molly at T'ai Chi.

STORE ROOM
7

SUNSHINE CAMP

Now we find ourselves building our own traditions against the background of a past summer camp where decades of little kids from the city came for a couple of weeks to play Indian, swim in the river and listen to ghost stories around the campfire. Some of them visit occasionally, all grown up now, and stand around grinning and remembering. Our friends are reminded of summer camps in their childhoods in other parts of the country. Even those who never went to camp are given the breath of an imagined scene never experienced until now. On a warm summer day, in the midst of outdoor tasks, I often stand transfixed by a spontaneous cessation of responsibility and care, a little boy again with dust in his nose, trees overhead and a blessed summer vacation stretching before him.

—Roland Jacopetti

Softball in the quad.

Breakfast with the Jacopettis.

Roland in the offal office.

Office space from a rescued shipping container.

Rosana at the kiln.

Institutional dimensions give
room for Alexandra's high warp loom.

Kelly in his rescued Signal Corps radar van, now a film animation studio.

Kelly's cabin.

George at the wok.

Hobart in his bike shop.

Hot tub.

AFTERWORD

Now that a new Ice Age cometh (dogs are sticking to the sidewalk) and fuel becomes more precious than virtue, I've come to realize they built the Bloomfield School facing in the wrong direction. Its twelve huge (four by eight feet) windows face west, toward the town of Bloomfield and incoming storms from the sea. If there were just some way of picking up the building and rotating it so the windows faced south, then we could enjoy the sun's heat and light throughout the day. But since this is beyond our means, I've devised other plans.

Two winters ago I had a bitter quarrel with an LP gas company which left me determined to design and build a new heating system for our schoolhouse that would minimize dependence on fossil fuels. The present system consists of two large LP forced air furnaces, one in each main room. They serve well, but suck gas like an old Cadillac and are ugly to behold. They both vent into sound brick chimneys that originally exhausted woodburning stoves. Wood storage rooms still adjoin these chimneys and I plan to reactivate them. My natural inclination for flea-market and thrift store shopping will provide what hardware I need for my improvised system, with wood and the sun providing the primary fuels.

First, I'll install three steel-framed storm windows in the blank south wall. These windows were donated and two woodworker friends volunteered their expertise. Cost: almost nothing; environmental impact: all day sunlight in the studio area (soon to be our living quarters). We'll put them in six feet high, just above the blackboards. Next I'll sell the gas heater, then clean and repair the chimney to accept the major investment required by the system I envision—a Fisher wood stove with water-heating copper pipes. This beautifully efficient, fire-brick lined heater can hold a fire for 24 hours. The system will also contain its own hot water storage tank, an old gas one with defunct apparatus, that Roland gave me. This is destined for the attic above the bathroom and will be connected to the gas water heater downstairs. Thus, by connecting the tanks in series, the heater's thermostat will regulate the whole system. When the wood stove is cherry red and supplying plenty of hot water, the gas heater will stay on pilot. When wood isn't burning the heater will pop on.

The entire system (a Five Year Plan) includes a solar greenhouse built against the south wall from old windows and will enclose the three new storm windows. They will then become symbiotic connectors between indoors and greenhouse, sun and heat from the greenhouse flowing into the house through a row of portals. On overcast days and at night the woodstove output will help keep the greenhouse warm. After the greenhouse is built, I'll install solar panels on the south slope of the roof—flat plate collectors containing copper tubing to feed into the hot water system. Since we have little freezing weather here, I hope to use an open system with potable water, rather than a closed system loaded with anti-freeze. The crowning glory of this Five Year Plan will be a combination hot tub and heat storage tank. A family with an abundance of hot water is wealthy indeed.

Though my plans are still sketchy, my engineering skills nil, and my budget tight while I continue to hustle and raise food for a family of five, I am still determined one day to emancipate myself from gas companies. After that, I'll become a carpenter again—convert the attic into bedrooms, build a darkroom, and on and on.

Nearly everyone we interviewed in the course of researching this book said, "Oh, I wish you'd come back next year when we have it more together." Buying a ready-made house is one thing—you simply move in and live there. But when you set out to rescue an old building, you embark on an affair of restoration and improvisation that can keep you busy for the rest of your life. With luck, you may even rescue yourself in the process.

—Ben VanMeter

Breakfast time in Bloomfield schoolyard.

Authors Ben VanMeter, Roland Jacopetti, and photographer Wayne McCall.